Y0-DWO-371

WORLD OF WONDER

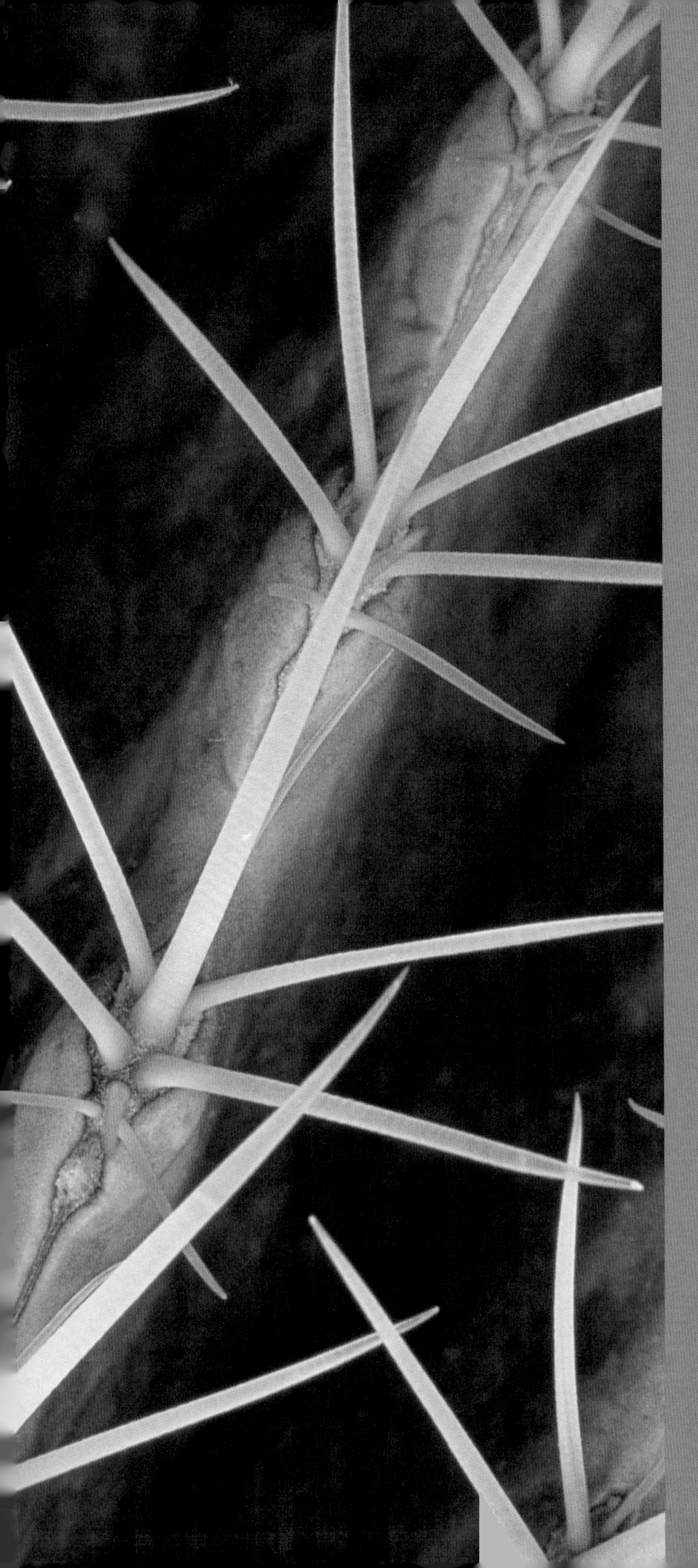

Published by Creative Education
123 South Broad Street
Mankato, Minnesota 56001

Creative Education is an imprint of
The Creative Company.

Art direction by Rita Marshall
Design by The Design Lab
Photographs by Affordable Photo Stock (Francis & Donna Caldwell), Peter Arnold (James L. Amos), Corbis (Carol Cohen, Michael & Patricia Fogden, Kevin R. Morris), Eyewire, Innerspace Visions (Doug Perrine), JLM Visuals (Breck P. Kent, Peggy Morsch), KAC Productions (Kathy Adams Clark, Larry Ditto, Bill Draker, John & Gloria Tveten), Root Resources (Mary & Lloyd McCarthy), James P. Rowan, Tom Stack & Associates (David Pleetham, Spencer Swanger)

Library of Congress Cataloging-in-Publication Data

Hoff, Mary King.
Handling heat / by Mary Hoff.
p. cm. – (World of wonder)
Summary: Describes the adaptations developed by various plants and animals to survive in very hot climates.
ISBN 1-58341-240-9
1. Acclimatization–Juvenile literature. 2. Adaptation (Biology)–Juvenile literature. [1. Acclimatization. 2. Adaptation (Biology)] I. Title.

QH543.2 H65 2002
578.4'2–dc21 2001047888

First Edition

9 8 7 6 5 4 3 2 1

cover & page 1: an iguana
page 2: cactus spines
page 3: a lizard on a cactus

Creative Education presents

WORLD OF WONDER

HANDLING HEAT

BY MARY HOFF

Plants with long roots that reach far beneath the soil Birds that pant to keep themselves cool Bacteria that love their habitat hot The world is full of creatures with special traits that help them tolerate hot, dry times and places.

SOME OF THESE adaptations keep creatures from getting so hot that they die. Others keep them from losing too much water or help them obtain water, which can be in short supply when temperatures soar. Either way, they help living things survive and pass life along to another generation.

This bird has adapted to life in the desert

SWEAT AND SPIT

When it's very hot, people begin to sweat. Moisture oozes from special glands to the surface of the skin. This is helpful because it takes heat to **evaporate** water. As the sweat evaporates from the skin, it draws heat away from the body, helping to cool it. A number of other creatures also sweat to stay cool, including horses and camels.

NATURE NOTE: *The hottest temperature ever recorded on Earth is believed to be 136 °F (58 °C), in Libya in 1922.*

Horse sweat is sometimes called lather

A waxy coating keeps cacti from drying out

Living things take advantage of the cooling properties of evaporation in other ways, too. Honeybees sprinkle water around their hives, using evaporation to cool the hives down. Kangaroos lick their front feet and let the saliva dry to cool off. Storks urinate on their legs so evaporation can cool them. Dogs, cats, and birds **pant**, exposing the moisture inside their mouths to the air. As the saliva evaporates, their bodies lose heat and are cooled.

Birds don't sweat, but they

do pant. Doves, pelicans, owls, and some other birds also have a special technique for using the cooling powers of evaporation called **gular fluttering**. When the bird gets very hot, it moves part of its throat rapidly, forcing air to move around inside the throat. As the air moves, it evaporates some of the moisture inside the bird, helping to cool down the bird's body.

NATURE NOTE: *Birds have no sweat glands, so they can't sweat to cool themselves. However, they do lose a lot of heat directly through their skin.*

Great blue herons pant to beat the heat

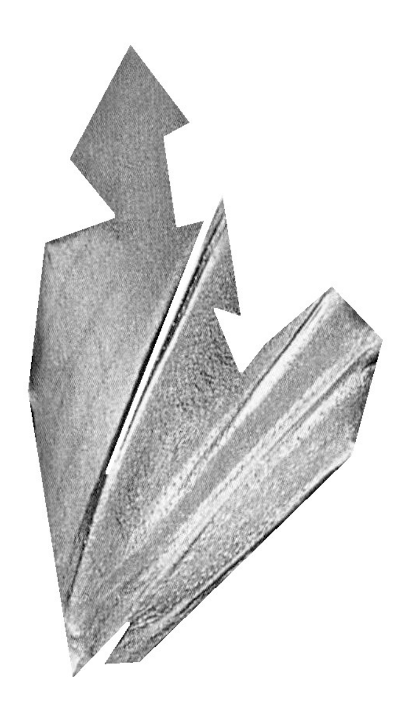

NATURE NOTE: *Australian fruit bats cool off by licking their wings and letting the saliva evaporate.*

OUTSIDES THAT COOL

When we think of the term "insulation," we usually think of protection from cold. But outside layers can insulate animals from heat, too. Many desert animals have light-colored skin, feathers, or fur. Since light colors absorb less light than dark ones, this coloration helps the animals to stay cool.

Some animals that live in hot habitats have dark fur or feathers. How does this help keep them from overheating? By absorbing the light, the dark layer prevents it from traveling to the animal's skin and heating it up. The heat stays on the fur or feathers, where breezes can blow it away.

♣ Plants have outsides adapted to tolerate heat, too. Many desert plants have tiny leaves, which absorb heat more slowly and release it more quickly than large leaves do. Light-colored stems and leaves reflect the sun's rays, keeping them from heating the plant too much.

NATURE NOTE: *Ostriches have been observed to survive temperatures of up to 133 °F (56 °C) by panting to cool themselves.*

Retama is a desert plant with tiny leaves

RADIATE-EARS

The kit fox, a cat-sized animal that lives in the deserts of North America, has big ears. These ears have a lot of surface area from which heat from the fox's hot blood can radiate, or move to the surrounding air. Jackrabbits, African elephants, fennecs (a kind of fox), and little mouselike creatures called jerboas lose heat through big ears, too. In goats and some other animals, horns and antlers with many blood vessels help to remove excess heat from the creature's body.

Since penguins are generally thought of as cold-climate animals, it's hard to think of them getting too hot. But penguins that live near the equator sometimes need to cool themselves. One way they do this is by holding their flippers away from their bodies to expose more of their skin to the cooling air.

NATURE NOTE: *A hot penguin will hold its flippers out away from its body to speed heat loss.*

HIDING FROM HEAT

Where is the coolest spot in a desert? Beneath the ground! Many animals burrow into the ground and spend the hottest part of the day there. Desert rats, geckos, rattlesnakes, kit foxes, and iguanas are among the animals that take advantage of the shade and cool soil they find beneath the surface of the earth. Some of these animals are nocturnal, which means they are most active at night. Others are crepuscular, or most active at dusk or dawn.

Some small birds avoid flying during the hottest time of the day to stay cool. Vultures, large desert birds, do the opposite. When it is very hot, they soar at high altitudes, where the air temperature is cooler.

NATURE NOTE: *The fennec and kit fox, known for their big ears, both belong to the dog family.*

Snakes called sidewinder adders hunt at dawn

TIME OUT

The desert tortoise, found in the Mojave Desert in California, goes underground in June and doesn't come out until fall. It goes into a long period of inactivity similar to hibernation. This state, called **estivation**, helps it to survive the intense heat and dryness. Ground squirrels, some toads, and other hot-climate animals also estivate.

Plants in hot places take time out, too, to survive the hottest, driest times of the year. Some make it through the heat in the form of seeds. Others retreat to bulbs beneath the ground. Still others, such as the ocotillo, stay alive above ground but become **dormant** until conditions improve.

NATURE NOTE: *Some seabirds wade in the water to stay cool. Heat escapes their bodies into the water through the skin on their legs.*

The desert tortoise burrows to escape heat

NATURE NOTE: *The roots of mesquite trees can extend more than 80 feet (24 m) down into the ground.*

BODY MOVES

The sand-diving lizard, which lives in the Namib Desert of Africa, tries to keep its toes cool by lifting its feet, two at a time, from the hot sand. When some lizards run across the desert floor, they run on two feet instead of four. Scientists think this helps them stay cool by reducing the amount of contact they have with the surface of the earth, which can be much hotter than the air because of the strong sunlight shining on it.

Cataglyphis bicolor, a Sahara

This lizard is raising two feet to cool them

Desert ant, stays cool in part because of its long legs. The legs hold the ant's body high above the hot surface of the earth. Some other desert insects also have long legs that help them avoid absorbing too much heat from the ground beneath them.

Some animals change their body position during the day to minimize the amount of heat they pick up from the sun. Camels and gulls are two animals that turn their bodies so only a small amount of their surface is hit by sunlight.

NATURE NOTE: *The ant* Cataglyphis bicolor *is a champion heat survivor. It has been known to survive temperatures above 140 °F (60 °C)!*

Long legs elevate desert ants above hot sand

WATER WAYS

Evaporation helps many creatures beat the heat by giving them a means to cool down. But evaporation can also be an enemy. If a living thing uses too much water to cool itself, it can become **dehydrated** and die. To survive, living things in many hot places have to not only handle heat, but also find ways to obtain and conserve the water they need to stay alive.

Different creatures have different ways of doing this. Some desert plants have roots that

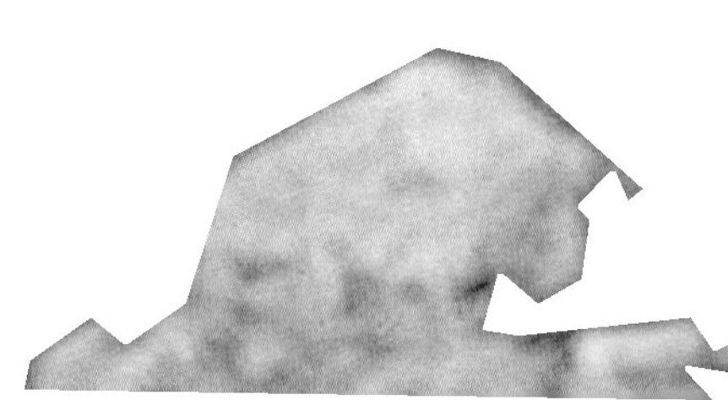

African mammals need watering holes to survive

reach far down into the ground where water can be found. Kangaroo rats, which live in the southwestern United States, make their own water by combining oxygen from the air they breathe with hydrogen found in the food they eat.

The saguaro cactus looks like it has no leaves at all, but in fact it has many leaves. They just are so skinny that they look like needles. Because these spines have less surface area than broad, flat leaves, they lose less water to evaporation.

Photosynthesis, the process by which plants use sunlight to make food,

Cacti store water as a jelly-like substance

The kangaroo rat makes its own water

requires water. Many desert plants use a special kind of photosynthesis that needs less water. Some desert plants also cope with the water problem by storing a lot of water inside their stems. They may also be covered by a layer of waxy material that slows the process of evaporation.

Camels have several adaptations that help them save

NATURE NOTE: *Plants with special adaptations to live in a hot, dry environment are called xerophytes, from the Greek words for "dry plant."*

Ocotillo plants grow leaves only when it rains

water. Their noses trap moisture that otherwise would escape in their breath. Camels also have the ability to let their body temperature become higher in very hot weather. This reduces the need to lose water by sweating. Kangaroo rats, zebra finches, and other animals that live in **arid** places also conserve water by concentrating their body wastes.

NATURE NOTE: *Lizards that live in hot, dry places are covered by a tough coating called keratin that helps to hold water inside their bodies.*

Camels can survive many days without water

HEAT LOVERS

Far beneath the surface of the ocean, super-hot liquids ooze up from inside the earth. Though this place is very inhospitable to most living things, creatures called **thermophiles** and **hyperthermophiles** call it home. One heat-loving bacterium found at the bottom of the ocean grows in an environment as hot as 235 °F (113 °C). In fact, if the temperature around it drops below 194 °F (90 °C), it stops growing.

Thermophilic bacteria are

Lava's heat can stimulate thermophile growth

also found in pools of hot water in Yellowstone National Park in the western United States. They create the red, yellow, and brown colors found in the pools. Thermophiles, including thermophilic fungi, are also found in compost heaps—piles of dead plants that are hot because they are in the process of decomposing.

NATURE NOTE: *Scientists use chemicals from thermophilic bacteria in* **genetic engineering***. The chemicals are valuable because they don't break down at the high temperatures the process requires.*

Yellowstone has a lot of hot underground water

Some bacteria live in geysers and hot springs

SURVIVING AND THRIVING

From panting pelicans to long-legged ants, the world's creatures have an amazing array of adaptations that help them survive high temperatures. These adaptations make it possible for them to survive and thrive in some of the most challenging environments on Earth.

Today, however, many of these animals and plants face a new challenge: alterations to their habitat brought about by humans. When we try to change parts of the world to suit them to our needs, sometimes we end up making it worse for the creatures that were adapted to the way things were. By considering the impact our actions have on the environment and its wild creatures, we can help ensure the future health and beauty of this amazing world, this world of wonder.

NATURE NOTE: *The word thermophile is a combination of the Greek words for "heat" and "loving."*

WORDS TO KNOW

Traits that help a living thing survive or reproduce under the particular conditions in which it lives are called **adaptations**.

Arid *places are very dry.*

An animal becomes **dehydrated** *when its body loses too much water.*

When a plant becomes **dormant**, *it temporarily stops growing.*

Estivation *is a period of prolonged inactivity that some animals go into during hot, dry times.*

Water absorbs heat when it **evaporates**, *or turns from a liquid to a gas.*

Genetic engineering *is the process of altering the DNA found in living things to make a new product or produce other benefits for people.*

Gular fluttering *is a fast movement of the throat used by birds to cool themselves by speeding evaporation of water.*

Hyperthermophiles *are creatures that thrive at temperatures above 176 °F (80 °C).*

When animals **pant**, *they breathe in and out very quickly. Some of the moisture in their mouths evaporates, cooling them.*

Thermophiles *are creatures that are adapted to live in very hot places.*

INDEX